ABOUT

TIME

TOO

ABOUT

··

TIME

··

TOO

···························

A Miscellany of Time

ROYAL
OBSERVATORY
GREENWICH

First published in 2021 by the National Maritime Museum, Park Row, Greenwich, London SE10 9NF

ISBN: 978-1-906367-66-4

At the heart of the UNESCO World Heritage Site of Maritime Greenwich are the four world-class attractions of Royal Museums Greenwich – the National Maritime Museum, the Royal Observatory, the Queen's House and *Cutty Sark*.

www.rmg.co.uk

A CIP catalogue record for this book is available from the British Library.

Design by Matt Windsor.
Typesetting by Thomas Bohm, User Design, Illustration and Typesetting.
Printed and bound in Spain by GRAFO.

10 9 8 7 6 5 4 3 2 1

WHAT IS TIME?

Time is something which affects us all in many different ways. It also generates some of the most intriguing questions asked by visitors to the Royal Observatory in Greenwich, the 'Home of Time'.

About Time Too provides the answers to these questions and also presents a whole range of other amazing facts and figures which show the influence of time on our daily lives. So why not take some time to sit back and find out more?

AGE OF THE EARTH

One way to visualise the passing of time is by imagining historical events as points on a very long ruler. For instance, it is possible to plot the history of the Earth, about 4.6 billion years, against an old English yard (91.5 centimetres). According to English folklore, the length of a yard equalled the distance between the tip of King Henry I's nose to his outstretched fingers. If the King were to sweep his nail file just once across the nail of his middle finger, then the amount of dust collected in the palm of his hand would represent the whole of human history.

AHEAD OF ITS TIME

What connects an 18th-century clock and a toaster? An amazing invention... called the

bimetallic strip. Clockmaker John Harrison created the revolutionary device to compensate for temperature changes in one of his famous marine timekeepers, known as H3. Harrison largely dismissed H3 at the time. After 19 years of painstaking work on the timekeeper, it still struggled to keep time as accurately as he wanted it to. Little did he know how versatile and useful his invention was! It would come into its own in the 20th century as it is now used as a thermostatic control in central heating, electric kettles, irons, toasters and more.

ALIENS IN SHELL SUITS

Radio and television signals travel at the speed of light, 186,000 miles per second. If an alien civilisation located 25 light years away from the Earth had the technology to receive these signals, they would now be able to watch television broadcasts that were made on Earth 25 years ago. Imagine aliens believing the Earth is inhabited by people like the Spice Girls or the Fresh Prince of Bel-Air, with hair like Rachel from *Friends* or dressed in shell suits! Perhaps the aliens are now enjoying their own late nineties revival!

ART OF TIME

Time is shown in many different ways around the world. In the East, it is represented as a dragon, the creative force of the Universe. The ancient Aztecs thought time was a snake with a head at both ends looking in opposite directions: one to the past and the other to the future. In the West, time is often symbolised as the well-known figure of Father Time (see page 32). 'Eternity' is often shown as a snake swallowing its tail which symbolises the never-ending cycle of time.

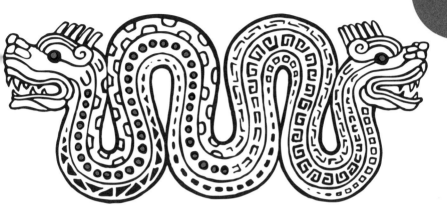

BACK TO THE FUTURE?

Some scientists believe there are a number of possible futures. If these many futures do exist, one way to get from one to another would be by travelling backwards in time and making different choices that then affect the future. This is known as the 'butterfly effect'. Films such as *Back to the Future*, *13 Going on 30* and *Harry Potter and the Prisoner of Azkaban* look at some of the devastating or drastic effects of meddling with the past. Some fictional stories explore the theory that these different futures run parallel to each other, meaning you can jump from one parallel universe to another. The Doctor and his companions often explore parallel worlds in the TV show *Doctor Who*, although with infinite potential futures it can be a tricky business finding the one you want!

BIG BANG THEORY

The Universe is believed to have originated about 15 billion years ago with a super-dense, super-hot globule of gas which expanded very rapidly. This is known as the 'Big Bang'. By the time the Universe was one second old, it had cooled to ten billion Kelvin, which is a thousand times hotter than the core of our sun. Nothing existed before the Big Bang: no energy, no space, no time. Scientists know what happened after the Big Bang, but it is impossible to say what happened fractions of a second before, because 'before' did not exist.

Ever since the Big Bang the Universe has been expanding and changing. When a star goes supernova its core collapses and the force of the resulting explosion blows the outer layer away. This outer layer then becomes the birth-place for new stars.

BIG BEN

The most famous clock in the world towers above the Houses of Parliament in London. Most people know the building as 'Big Ben', but it's actually called the Elizabeth Tower, named after the Queen in celebration of her Diamond Jubilee. 'Ben' is the name of the 13-tonne bell which strikes the hours. It may be named after the man who commissioned it in 1856, Sir Benjamin Hall. There have actually been two Big Ben bells. The first one cracked and was melted down and recast in 1858. The current bell also cracked soon after it was installed, but still manages to do the job. The chimes of Big Ben are iconic across the world, but they haven't been heard for a while. The bell has been undergoing major repairs since 2017, although it still chimes for big events such as New Year's Eve and Remembrance Day thanks to a temporary mechanism. Politician Boris Johnson floated the idea of raising '£50,000 a bong' to make the bell chime for Brexit on 31 January 2020, marking the exact moment that the UK officially left the EU. However there wasn't enough time to make it happen, although several crowdfunding pages attempted to raise some of the cash for the famous bongs.

BITTEN BY THE (MILLENNIUM) BUG

The year 2000 will be remembered as the year the world celebrated twelve months too soon. Popular opinion favours the numeric elegance of 2000 as the year to celebrate the dawning of the third millennium, whereas it really signified the waning of the second millennium. It also signified a panic as some people believed the 'millennium bug' would cause the end of the world. It was thought that computer programs, or anything with a chip, would crash when the clock ticked over to 00.00 on 1 January 2000. The anticipated Y2K problem led to $300 billion of computer upgrades worldwide but it turned out to be a false alarm.

THE BLINK OF AN EYE

On average, adults blink around 15–20 times per minute – that's anywhere up to 28,800 blinks a day. It can be as low as three or four times per minute when your eyes are focused on something, like when reading a book. Since each blink takes approximately a quarter of a second, about an hour of our waking day is spent with our eyes partly or completely closed.

A BRIEF HISTORY OF TIME

TIME magazine gained its timely name as it was designed to be a quick read. The creators, Briton Hadden and Henry Luce, wanted to create something that anyone could read in less than an hour. TIME is famous for its Person of the Year feature where it recognises people for their influence on the year's events... 'for better or for worse'! The eclectic list of 'winners' includes Greta Thunberg, Barack Obama, Adolf Hitler and The Computer. Franklin D. Roosevelt has gained the title the most times, in 1932, 1934 and 1941.

BRIGHT NEW DAY

For us, every new day begins at midnight, but this has not always been the case. Astronomers often measured the start of the day from 12 noon so they did not have to change dates during the night when they were busy observing the stars. In Ethiopia, people begin their day at dawn. The ancient

Hebrews measured the beginning of the day from sunset, a custom which is still observed in Jewish festivals today.

BUGS WITH BODY CLOCKS

Some bacteria have biological clocks similar to those found in human beings. Cyanobacteria are the most primitive organisms known to have a daily cycle of photosynthesis, the process by which a plant converts sunlight into energy to help it grow. Even when they are kept in darkness without light from the Sun, these bugs automatically 'switch on' for photosynthesis as if it were daytime.

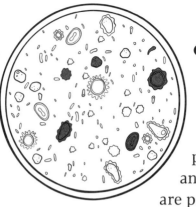

CALENDARS GALORE

Calendars help us to keep track of time over long periods by marking the passage of days, months and years. These units of time are primarily based on natural astronomical cycles. A day is the time it takes the Earth to complete one full rotation on its axis. A month is based on the time it takes for the Moon to orbit around the Earth, and a year is based on the time taken for the Earth to orbit around the Sun. The problem for calendar makers is that none of these natural cycles fits in exactly with the others. There is no simple solution to the problem, as indicated by the fact that there have been many different calendars used throughout history and across cultures.

CHINESE CALENDAR

Chinese years are named after 12 animals, which follow one another in rotation: the rat, ox, tiger, rabbit, dragon, snake, horse, goat, monkey, rooster, dog and pig. The Chinese New Year is celebrated on the second New Moon after the winter solstice (see page 135). This can fall anywhere between 21 January and 19 February. You can find the Chinese year in which you were born by checking the chart on the next page.

Some people are born pigs!

Rat	1948	1960	1972	1984	1996	2008	2020
Ox	1949	1961	1973	1985	1997	2009	2021
Tiger	1950	1962	1974	1986	1998	2010	2022
Rabbit	1951	1963	1975	1987	1999	2011	2023
Dragon	1952	1964	1976	1988	2000	2012	2024
Snake	1953	1965	1977	1989	2001	2013	2025
Horse	1954	1966	1978	1990	2002	2014	2026
Goat	1955	1967	1979	1991	2003	2015	2027
Monkey	1956	1968	1980	1992	2004	2016	2028
Rooster	1957	1969	1981	1993	2005	2017	2029
Dog	1958	1970	1982	1994	2006	2018	2030
Pig	1959	1971	1983	1995	2007	2019	2031

CLOCKING IN

Many people think that the practice of 'clocking in' and 'clocking out' was introduced in the 19th century with the Industrial Revolution. However, it was actually used much earlier by the ancient Egyptians, when the pyramids were being built. There are still surviving Egyptian records that list the days people worked. Modern clocking-in systems record the exact hours people work in order to calculate wages. They may be used for catching people who arrive late or leave early and they record the number of work hours lost due to illness. One study estimated that the common cold costs the US economy more than $40 billion with 189 million working days missed due to colds.

CLOCKS

The word 'clock' is taken from the French word '*cloche*', meaning a 'bell' or 'gong'. The modern word is derived from the fact that the earliest clocks simply marked the passing hours by striking a bell. These clocks were used in monasteries, allowing the monks to observe the many prayer times in their strictly regulated day.

DANDELION CLOCKS

The downy head of the dandelion, which contains the seeds of the plant, is traditionally used by children as a game for guessing the time. By counting the number of puffs it takes to blow all the seeds off the stalk, children learn to recite the hours of the day. The game also helps to maintain the life cycle of the dandelion, by scattering the seeds and improving the plant's chances of sprouting the following year.

DATES FOR THE WORLD

The Gregorian calendar is used throughout the world even though it is essentially a Christian version of an ancient Roman calendar. Many non-Christians use the Gregorian calendar but they also have their own calendars which reflect a yearly cycle of their own religious beliefs and festivals. Whereas the Gregorian system uses the phrase 'Anno Domini' (AD) to date its years, Jewish people use the phrase 'Anno Mundi', meaning 'the year of the world'. This system is based on a belief that the world was created on what Christians would call 7 October 3761 BC. Muslims use 'Anno Hegirae', counting their years from the Prophet Muhammad's emigration (Hegirae) from Mecca to Medina on 16 July AD 622. In an attempt to replace the Christian terms BC and AD, some people use the abbreviations 'BCE' (meaning 'before the common era') and 'CE' ('the common era').

THE DAYS GROW LONGER

Over the last 2,700 years, the length of a day has gradually increased at the rate of 0.0015 seconds per century. This is due to the gradual slowing down of the daily rotation of the Earth on its axis caused by the Moon's tidal pull on our oceans. Melting glaciers, earthquakes, strong winds, ocean currents and disruptions in the Earth's core can also produce short-term effects on our planet's spin.

DECIMAL TIME

We are all familiar with having 24 hours in a day, 60 minutes to an hour and 60 seconds to a minute. This timekeeping system is known as 'duodecimal', which means 'twelve-based'. After the revolution in France in 1789, it was decided to introduce a 'decimal' (or 'ten-based') system of measuring time, where the day was divided into ten hours, each having 100 minutes and each minute having 100 seconds. Many curious clocks and watches were produced at this time, usually with two dials to give both the old style and the new style time. These are now very rare as mandatory use was quickly phased out. It was unpopular primarily because it made all transactions with other countries, still using a 12-based time system and calendar, very difficult.

A DECK OF WEEKS

The traditional pack of 52 playing cards, introduced to Europe during the 15th century, symbolises the calendar. The 52 cards represent the number of weeks in a year;

the four suits (which we now know as clubs, spades, diamonds and hearts) represent the four seasons; the thirteen cards in each suit represent the number of weeks in a season; and the red and black cards represent day and night.

DIRECTION OF TIME

In many countries and cultures, time and direction are closely linked. For example, knowing the time and direction is very important for Muslims. As well as needing to know the correct times to pray, it is important for Muslims to know which direction to face for Mecca. Many indigenous peoples of the Americas, including the ancient Inca and the Hopi, use the horizon for telling the time. As the Sun and the stars appear to rise and set at slightly different points on the horizon throughout the year, it is relatively easy to establish a set of markers on the distant skyline which can be used to tell the date.

DOING TIME

The phrase 'doing time' originates from Australia and means 'to be in prison'. Prisons in New South Wales were run on a strict time system and this even began to affect their architecture. Numbered cells were arranged around the circumference of the prison yard like the minute intervals on a gigantic clock.

THE DOOMSDAY CLOCK

There's a rather sinister imaginary clock counting down to the end of the world. Midnight represents the apocalypse on the

Doomsday Clock, founded by the Bulletin of the Atomic Scientists. It's adjusted regularly to convey the threats facing humanity and the planet, one of the most pressing being climate change. Unfortunately the minute hand was set at 100 seconds to midnight in January 2020, the first time it has been kept in seconds rather than minutes.

EASTER EGGS

The egg is a symbol of fertility and rebirth. It ties in quite naturally with the Christian festival of Easter which commemorates the resurrection of Jesus Christ after his crucifixion. The name 'Easter' originates from the Anglo-Saxon pagan goddess of dawn, Eostre. As dawn signifies the rebirth of the day, Easter is associated with the return of spring. It is actually possible to celebrate Easter twice in one year. The Orthodox church calculates the date using the old-style Julian calendar while the Western church uses the modern Gregorian calendar. As a consequence, both churches rarely celebrate Easter on the same day.

ECLIPSED BY THE MOON

If you stand in one place on the surface of Earth for long enough, you will eventually witness a total eclipse of the Sun. A solar eclipse happens when the Moon passes in front of the Sun, casting a shadow on Earth. During a total

eclipse, the day seems to turn into night for a few seconds or even several minutes. A total solar eclipse is visible from Earth on average every 18 months. However, it takes about 45 centuries for the whole world to be covered by the tracks of solar eclipses.

THE END OF TIME

Scientists believe that time began with the Big Bang, a great explosion which caused the universe to be created and expand outwards. However, scientists do not agree on how the universe will end. Some believe that it will eventually stop expanding and start to contract, ultimately resulting in what is known as the Big Crunch. Others say this will not happen and instead the Universe will expand forever. The real question, as yet unanswered, is this: if the Universe begins to contract, will time then start to run backwards?

FAST FOOD

When Clarence Birdseye (1886–1956) began working as a naturalist near the Arctic, he made a chilling discovery. He noticed that a freshly caught fish, when placed on the Arctic ice, froze solid almost immediately and that it was still fresh when thawed much later. The freezing process kept the food 'suspended in time', preventing it from deteriorating. In 1922, Clarence set up his own company, Birdseye Seafoods, Inc., which is still known today for frozen peas and fish fingers.

FATHER TIME

The image of Father Time is a familiar way of depicting the idea of 'time' in human form. He is often shown as an old man with symbols connected with the passing of time, such as a sandglass with the last few grains running out, a snuffed candle, a skull and the reaper's scythe. The message is clear: 'time' creates and destroys all things and nothing can escape.

FIRE ALARMS

Candles and incense sticks burn at an even rate and can be used to measure intervals of time. Early alarm clocks would have strings with weights hung over a long stick of incense. As the incense smouldered, it would burn through the strings and the weights would drop in a tray below, marking the time with a loud clatter. In ancient China, long-distance messengers would use a stick of incense as a slightly more dangerous alarm clock. As they always needed to be ready to deliver urgent messages, they often only had time for quick naps. To ensure that they never slept too deeply, they would light a short stick of incense and place it between their toes. As the burning tip reached the skin, the messengers would wake with a start!

FLOW OF TIME

A sandglass or hourglass measures intervals of time according to the flow of fine particles. The name is misleading because these glasses can measure any period from a few seconds up to several hours and may be filled with any granulated substance. Sand, ground eggshells and even ground human bones have been used to fill them!

FLOWER POWER

Can you grow your own clock? Gardeners have long known that certain flowers tend to open or close at specific times during the day. The idea of a floral clock was first put forward by the 18th-century Swedish botanist Carolus Linnaeus (1707–78). They may not be as accurate as an atomic clock but they are certainly prettier!

FLYING TIME

Birds and butterflies in northern Europe migrate south for the winter to avoid harsh weather and find new supplies of food. They know the date by sensing changes in temperature from summer to winter and the differing amounts of daylight. Global warming has meant that spring weather is arriving earlier in the year, with trees and flowers blossoming earlier today than they did in the 1950s. For every single degree rise in temperature, swallows arrive on our shores approximately three days earlier.

FREAKY FRIDAY

Friday has long been regarded as an unlucky day by superstitious people, probably because

it was the usual day for executions. It was often known as 'hangman's day' and people are called 'Friday-faced' if they look sad. The number 13 has long been related to bad luck. The irrational fear of the number 13 even has a name – triskaidekaphobia. Some people believe it is unlucky when the 13th day of the month falls on a Friday. Any month that begins on a Sunday will have a Friday 13th and every year must have at least one Friday 13th but never more than three – a good puzzle to work out! When Friday 13th occurs three times in a year, some people think it predicts a major catastrophe.

FROZEN IN TIME

Disastrous events have sometimes been frozen forever in time, as a grisly reminder to the future. The bodies of some of the men who accompanied Sir John Franklin on his doomed expedition to the North American Arctic in 1845 were found many years later, perfectly preserved, frozen in the permafrost. In a museum in Hiroshima, Japan, there are several watches which remain permanently stopped at 8.15 a.m., the moment on 6 August 1945 when the first atom bomb was detonated over the city. A watch was found on the body of one of the victims of the *Titanic* disaster. Robert Douglas Norman, a 27-year-old Scot, was a passenger on board the ship when it sank on 15 April 1912 and his watch is frozen at the time the ship sank, the hands being too rusty to continue moving after being plunged into the icy water.

GESUNDHEIT

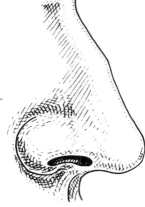

Or 'bless you' in German! Many scientists have tried to work out the speed of a sneeze. For a long time it was estimated to be around 100 mph! Some studies state it is actually nearer to 200 mph but one in 2013 found the maximum velocity to be just 10 mph. It may vary according to body size but either way, you might need a box of tissues!

GODS OF TIME

The ancient Egyptians believed in a sky goddess called Nut who gave birth to the Sun every morning and swallowed him every evening, creating a never-ending cycle of day and night. The Chinese have a god of longevity or long-life called Shou Lao, with a long bald head and often depicted holding symbols of eternity, such as a peach, and animals which are reputed to have long lives, such as the stork. The Aztecs believed that time was kept on course by offering human sacrifices to appease the sun god, Tonatiuh. These ensured that he would rise each day and cross the sky. Children were sacrificed in the months when rain was expected. The more they wept, the better the omen for rain.

GREENWICH MEANS TIME

Greenwich Mean Time (GMT) is a way of converting solar time, as measured by an observer at Greenwich, into clock time. The time interval between noon each day is not exactly 24 hours, it varies with the seasons. Astronomers created a consistent 'mean' time that was easy to use with a clock all year round. For many years, the Royal Observatory was responsible for distributing GMT to the nation. It was not until 1884, when it was agreed that the meridian passing through the principal telescope at Greenwich would represent the Prime Meridian, or longitude 0°, that GMT became the international time standard.

GREENWICH: THE 'HOME OF TIME'

The Royal Observatory, in Greenwich, was founded in 1675 by King Charles II, and from the beginning the Observatory was associated with accurate timekeeping. It was founded to improve navigation at sea and John Flamsteed was appointed as the first Astronomer Royal. His task was 'rectifying the tables of the motions of the heavens, and the places of the fixed stars, so as to find out the so much-desired longitude of places for perfecting the art of navigation'. Astronomers Royal worked on these star catalogues for many years to improve navigation, before later expanding to become a testing centre for the newly invented marine timekeepers in the 18th century. With the installation of the iconic Greenwich Time Ball in 1833, and the Greenwich Meridian being adopted as the Prime Meridian of the World in 1884, the Observatory has without doubt earned its place as the 'Home of Time'. The sea, the stars and time are inextricably linked, and the Royal Observatory still embodies this today.

GREENWICH TIME LADY

In 1892 Ruth Belville, the 'Green-wich Time Lady', took over her father's business of distributing GMT to a network of subscribers. Every Monday, Ruth would travel by public transport to Greenwich, have a cup of tea with the porter, receive a certificate of accuracy for her trusty chronometer, and spend her day travelling to London's clock shops, bringing her 40 customers the correct time, whereupon they would adjust their own clocks. Ruth carried on in this vein through the First World War and even when competition arose from the Standard Time Company and the breakthrough of wireless communication. She chose to retire in 1940, at the age of 85, having completed her rounds for 48 years!

GREGORIAN CALENDAR

By the 16th century, the Julian calendar had become out of step with the natural cycle of the seasons. The problem was that the Julian year was eleven minutes and fourteen seconds too long. This does not sound much but amounts to a whole day every 128 years. New calendar changes were needed and, in 1582, Pope Gregory XIII declared that ten days should be chopped from the current calendar to bring it back in line with the stars. The new calendar was known as the Gregorian calendar, after the Pope, and was quickly adopted in Catholic countries such as France, Italy and Spain. By the time the calendar was introduced into England in 1752, the calendar had slipped by an extra day. This meant that eleven days had to be dropped, and 2 September was followed the next day by 14 September. The Pope also introduced a new rule for calculating leap years. The Gregorian calendar is the one we still use today. Of course, it is not exactly correct but it should suffice, with very minor adjustments, for the next 3,000 years.

HAPPY DEATH DAY!

Many of us probably don't want to know how long we have left in this world. This didn't stop the creation of morbid Death Clock websites. After answering some simple questions including your age, BMI and smoker status, an algorithm calculates your own personal day of death! Of course, it's just a bit of fun and shouldn't be taken too seriously. You'll be pleased to hear that the global average life expectancy is currently over 70 years, rising to nearer 80 years in countries such as the UK, the US and Canada.

HARRISON, THE TIME LORD FROM YORKSHIRE

In 1759, the clockmaker John Harrison (1693–1776) invented the most important clock ever made. This marine timekeeper solved the 'longitude problem', the greatest scientific puzzle of the 18th century. Harrison was the first to develop a timekeeper which could cope with the motion and changing temperature and humidity on board a voyaging ship, and still keep accurate time. At last ships all over the world could keep Greenwich Mean Time and use it to establish their position on the featureless oceans. Captain James Cook took a copy of this watch on his voyages to the South Pacific and called it 'My trusty friend, the watch' because it was so reliable. Many of John Harrison's ground-breaking clocks are on display at the Royal Observatory, Greenwich.

HEAT OF THE NIGHT

More people die in the early hours of the morning than at any other time during the day. This is when body temperature is at its lowest, a tricky time for a failing system. Body temperature begins to rise as we approach the time at which we normally wake up. We are at our warmest in the late afternoon and early evening. Olympic records are most likely to be broken later in the day, when athletes' muscles are fully warmed up.

HIDDEN WEB

Even spiders have a part to play in the history of timekeeping. The Prime Meridian passes through the centre of the principal telescope, known as the Airy Transit Circle, at the Royal Observatory, Greenwich. This telescope was considered at the forefront of modern technology when it was built in 1848. Every day, astron-omers used the telescope to

measure the position of the Sun and the stars as they crossed over the meridian. They did this in order to establish accurate time. The only material that was strong, fine and sticky enough to form the cross-hairs of the telescope's eyepiece were the threads of a spider's web. The astronomers had to rely on the skills of the Royal Observatory gardener who looked after the spiders to ensure their high-tech piece of equipment could function properly.

IF THE RAIN'S GOT TO FALL

People often grumble that it always seems to rain at the weekends, but scientists now think that the complaint may be justified. The weekly work schedule of human beings seems to have an impact on the weather. Throughout the week, pollution and ozone levels rise steadily as factories and cars chug out fumes and dust particles which collect in the atmosphere. The particles cause clouds to form offshore. When people finish work on Friday evening, the pollution levels fall and the clouds move inland to give us a rainy weekend!

IMMORTAL ANIMALS

According to scientists, it is impossible to work out the age of certain species of animals due to some clever biological traits. Some types of jellyfish are able to reverse their life cycle rather than dying, making them 'biologically immortal'. Flatworms are almost immortal too as they can regenerate their cells, making them seem forever young. But sometimes the ages of animals can be difficult to verify for more practical reasons. Aldabra giant

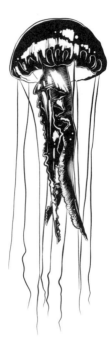

tortoises, who boast a potential lifespan of more than 200 years, have an annoying tendency to out-live the humans observing them!

IN A JIFFY

We often use the word 'jiffy' to show that something will be done quickly. The informal term has been around since the late 18th century. The earliest known use was in an edition of *Town and Country Magazine* published in 1780. But this isn't just a slang term, it's actually a specific length of time. The scientist Gilbert Newton Lewis (1875–1946) defined jiffy as 'the amount of time it takes light to travel one centimetre in a vacuum' – about 33.4 picoseconds. To you and me that's trillionths of a second, so by Gilbert's terms no-one has ever truly been 'back in a jiffy'!

IN THE LONG RUN

Marathons date back to Ancient Greece. The legend says that in 490 BC the Greek messenger Pheidippides was sent back from the Battle of Marathon to Athens to report the news of Persians' defeat, giving the 26.2-mile race its name today. Modern marathons have become faster and faster with current male and female professionals completing races in near to two hours! Eliud Kipchoge ran the first sub-two-hour marathon in Vienna in 2019. But his time of 1:59:40 wasn't officially recognised as the world record as the event wasn't an open competition and he used a team of rotating pacemakers. It did show the possibility of achieving the record time though and athletes are constantly pushing their bodies to the next level in sport. For those looking for the ultimate challenge, ultramarathon events now exist. You can run races

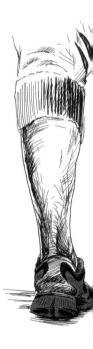

all over the world ranging from 30 miles to 1,000 miles, the latter taking more than 10 days to complete!

> **GLOBETROTTERS**
> *The fastest time to complete a marathon on each continent for a man is 6 days, 18 hours, 2 minutes and 11 seconds. It was achieved by Douglas Wilson from Australia from 17 January to 24 January 2015 as part of the 2015 World Marathon challenge. The record for the fastest time set by a woman is 6 days, 18 hours, 38 minutes and was achieved by Becca Pizzi from the USA.*

INTERNATIONAL DATE LINE

Imagine flying above the North Pole so the Earth looks like a giant sphere, which can be divided into 360°. The Prime Meridian marks 0°, the line from where all time and longitude are measured. On the opposite side of the circle, at 180°, is the International Date Line. The date is different on either side of the Date Line. When it is Monday to the west of the Line, it is still Sunday to the east. By crossing the International Date Line from west to east, it is possible to celebrate the New Year or a birthday twice. If the celebrations begin in a country to the west of the line (such as Australia, New Zealand or Japan), it would be possible to catch a flight just after midnight to the east, over the Line, to Hawaii, Samoa or the USA. By losing a day as you cross the Date Line, you will arrive just in time to celebrate again!

JET LAG

Modern travel on board jet aircraft means that we can 'cheat' time when we travel long distances. It was once an amazing achievement to travel around the world in 80 days. At that speed, the body clock could easily cope with resetting by about 20 minutes per day. Air travel can now deposit people at destinations with their body clocks up to 12 hours out of phase with local time. The body clock is designed for a regular rhythm of day and night. It is thrown out of phase when these rhythms are disturbed. This disturbance is what we call 'jet lag'. Symptoms of jet lag include tiredness and loss of concentration, which persist until the body clock adjusts to the new time. Those who

are used to a fixed daily routine are often the worst sufferers. Children under three years old do not usually suffer jet lag as they are more adaptable.

JEWISH CALENDAR AND PASSOVER

The Jewish calendar is dated from the year in which the Jews believe God created the world. It was calculated using biblical data and coincides with 3761 BC in the Gregorian calendar. The Jewish feast of Passover commemorates the events of the night which convinced Pharaoh to free the Jews from Egypt. It begins on the eve of the fifteenth day of the lunar month of Nisan.

JUST A MINUTE

The English words for minutes and seconds come from the Latin. The 'minute' was so-called because it is a small, or minute, fraction

of an hour. The 'second' is a secondary measurement, next in line to the minute. It is a
minute (tiny) part of a minute (in time).

JUST A SECOND

The second is a fundamental unit of time.
In practical terms, the second is 1/60 of a minute, 1/3,600 of an hour, or 1/86,400 of a day.
If you thought measuring time was simple,
try this scientific definition of a second for
size. In 1967, the second was defined as: '... the
duration of 9,192,631,770 cycles of radiation
corresponding to the transition between two
hyperfine levels of the ground state of the cae-
sium-133 atom'.

KETCHUP TIME

Heinz Tomato Ketchup has an official speed of
0.028 miles per hour as it leaves an upturned
bottle. This may seem slow but any faster and
it is rejected at the factory! This didn't stop
André Ortolf achieving the bizarre record of

'fastest time to drink a bottle of ketchup' in Germany in 2017. He drank a whole bottle (396g) in 17.53 seconds.

KILLING TIME

According to Guinness World Records, Iwao Hakamada is the world's longest-serving death row prisoner. He spent 45 years on death row in Japan before he was freed in March 2014, following suggestions that police investigators may have fabricated the evidence on which he was convicted. He spent much of that time in solitary confinement.

LEAP YEAR BABIES

Leap year babies are very special. A leap year contains 366 instead of the usual 365 days. The extra day is added to the end of February, which has 29 days in a leap year rather than 28. This means that anyone  born on 29 February normally has to wait four years for each birthday! But around the years 1700, 1800 and 1900, people born on 29 February had to suffer an eight-year gap between birthdays – see below for explanation!

LEAP YEAR RULES

The calendar has needed occasional changes to keep it synchronised with the stars. As it takes about 365 days for the Earth to make a complete orbit of the Sun, the quarter days are saved up and added all at once every fourth year. These are known as 'leap years' and they contain 366 days instead of the usual 365. To calculate a leap year, the year number should be exactly divisible by four. For example: 1848, 1996 and 2020 are all leap years. Even so, this rule creates three days too many in a 400-year period, so an exception to the rule was created by Pope Gregory XIII. In order for a centenary year (a year ending 00) to qualify as a leap year, it must be exactly divisible by 400, as with the years 2000 or 4000. This rule means that 1800 and 1900 were not leap years. Although they are divisible by four, they are not exactly divisible by 400.

LEAPING SECONDS

Just as we have leap years, we also have leap seconds. With the introduction of highly accurate atomic clocks, it was realised that the Earth and Sun are far from perfect timekeepers. In addition to this, it was discovered that the rotation of the Earth is gradually slowing down. To keep atomic time synchronised with Greenwich Mean Time (time measured by the Sun), an occasional extra second known as a 'leap second' is sometimes added to the year. The first leap second was added in 1972. The number of leap seconds that might be added changes every year, but the only time where they are added is either 31 December or 30 June at midnight GMT. At that moment, the six-pip time signal broadcast on BBC radio becomes a seven-pip signal. Since the leap second is always added at midnight GMT, this corresponds to it being added at 1 a.m. the following morning in Paris and 7 p.m. the previous evening in New York.

LIGHT YEARS AWAY

A light year measures the distance you could cover if you travelled at the speed of light, around 186,000 miles or 300,000 kilometres per second, for a whole year. A ray of light travels 5.88 trillion miles a year in space. The bright star Sirius is located nine light years away from Earth. Our galactic neighbour, the Andromeda Galaxy, is 2.5 million light years away, yet incredibly we can still see it with the naked eye! The famous physicist Albert Einstein (1879–1955) discovered that if you were to travel close to the speed of light, the time you experienced would be slower than for those who stayed behind on Earth. This means that, in theory, you could send your parents zooming off at around the speed of light for 40 Earth years and they would come back younger than you!

LINES OF TIME

Nature provides us with many clues to help tell the ages of animals and plants. The age of a

salmon can be counted by the number of lines on its jawbone and the age of a tree corresponds to rings which run in circles from the centre of its trunk. As horses grow older, their teeth grow longer. At the age of ten, they develop a groove in the tooth known as Galvayne's groove, which itself becomes longer as the horse grows older. This may have given rise to the phrase 'long in the tooth' which may refer to a very old person or animal.

LONG-PERIOD COMETS

How long would you have to wait to see the same comet twice? It depends on the comet! The famous Halley's Comet, named after the second Astronomer Royal Edmond Halley (1656–1742), is only visible to the naked eye once every 75 years, meaning you may only see it once or twice in your lifetime. It won't appear again until 2061 but that sighting is likely to go down better than the one in 1910. There was widespread panic when astronomer Camille Flammarion claimed the comet's poisonous tail would 'impregnate the atmosphere and possibly snuff out all life on the planet'. The announcement caused panic-buying of gas masks and anti-comet pills, but in reality the gas in the tail is so diffuse it has no effect when it passes through the Earth's atmosphere.

LOOKING BACK IN TIME

When we look through a telescope at the distant stars, we are really looking backwards in time. This is because it takes time for starlight to travel to the Earth. Light travels at about 186,000 miles per second. It takes about 8½ minutes for light from the Sun, our nearest star, to travel to the surface of the Earth. This means that the sunlight we see and feel is

already 8½ minutes old by the time it reaches us. This difference between the time at which we perceive light and the time it was created is known as 'lookback time'. Modern telescopes reveal galaxies so far away that the light reaching us now initially set out when the galaxy was forming. A recently discovered galaxy situated 12.3 billion light years away is so distant that astronomers are able to look back to the early universe, just 1.5 billion years after the Big Bang (see page 12).

MANY MONTHS

The word 'month' comes from the Moon and reflects the fact that most early calendars were based on the Moon's phases. About 29½ days elapse from new moon to new moon. The early Roman calendar contained ten months, beginning with March. From this calendar, we have retained the names for the months of March (after Mars, god of war), April (from the Latin, meaning 'budding'), May (after Maia, a very ancient Greek goddess), and June (from the Latin, meaning 'youth'). We have also retained the numerical names of September, October, November and December which are derived from the Latin words meaning 'seventh', 'eighth', 'ninth' and 'tenth'. Later on,

January (named after the two-faced Roman god, Janus) and February (after the Roman god Februus) were added to make a 12-month calendar. This pushed the numerical months out of order, so that September became the ninth month and December the twelfth. Two months were renamed – Quinctilis became

MARS.

July (after the Roman emperor, Julius Caesar) and Sextilis became August (after the first Roman emperor, Augustus).

MOVABLE FEASTS

A movable feast is a religious event which takes place on a different calendar date each year. For example, Easter is a movable feast in the Christian calendar. It takes place on the first Sunday after the first full moon on or following the spring equinox (the day when the hours of daylight and darkness are equal, around March 21). If the full moon occurs on a Sunday, then Easter will be the following Sunday. It can fall as early as 22 March or as late as 25 April.

MYSTIC SIXTY

The ancient Babylonians were geniuses when it came to maths. They believed that 60 was a mystical number and developed a system of counting to the base of 60, called 'the sexagesimal system'. Sixty was chosen for convenience

because it is divisible by a large number of smaller numbers without leaving a remainder. The numbers 2, 3, 4, 5, 6, 10, 12, 15, 20 and 30 all divide evenly into 60 without remainders. Our present time-keeping system, with its 24-hour day, 60-minute hours and 60-second minutes, reflects its Babylonian origins.

NATIONAL PHYSICAL LABORATORY

The National Physical Laboratory (NPL) at Teddington in Middlesex, is the home of the caesium fountain clock in the United Kingdom and the place where the latest developments in precision timekeeping are made. The scientists at the NPL produce the time signals that help us keep accurate time and they keep an eye on leap seconds. They are at the cutting edge of new technology, and are always working towards improving timekeeping standards.

NEW CENTURY,
NEW MILLENNIUM

At the end of the 19th century, there was a great deal of discussion over when the new century would actually begin. The seventh Astronomer Royal, Sir George Biddell Airy (1801–92), took part in the discussion and wrote a letter to *The Times* explaining that the official start of the new century would be the year 1901. The same debate erupted about the start date for the 21st century, which heralded the beginning of the third millennium. The official start date of the third millennium was 1 January 2001. As there is no year zero in our calendar, the sequence of years passes from 1 BC to AD 1 without a year zero. The first century began with the year AD 1. If one thousand is added to the year 1, it takes us up to year 1001, which marks the beginning of the second millennium. Adding another thousand years takes us to the start of the third millennium, or 2001. The next millennium will start on 1 January 3001.

NEW YEAR CELEBRATIONS AROUND THE WORLD

The way in which people celebrate the New Year changes from country to country. In Japan, debts must be settled and the house cleaned before a new year begins. Japanese New Year food is sea bream, a type of fish which is believed to bring good luck. In Spain, people count down the last minutes of the old year while popping grapes into their mouths. In the southern part of the United States, black-eyed peas are eaten on New Year's Day for good luck. In Hungary, a roast pig with a four-leaf clover in its mouth is prepared. In Greece, a cake called 'Vasilopita' is baked with a coin inside it. The person who gets the slice with the coin is thought to have special luck in the coming year.

The Jewish New Year is celebrated with apples dipped in honey. The Buddhist New Year is celebrated in Tibet with a soup called 'guthuk', which is served with dumplings filled with symbolic ingredients like salt, chilli and charcoal.

NEW YORK MINUTE

Ever heard of the phrase 'in a New York minute'? This measure of time is a quick one, often being used to mean that something gets done right away. It is said to have been coined in the late 1960s, not in New York itself but in Texas, the theory being that a New Yorker does in an instant what a Texan takes a minute to do. American entertainer Johnny Carson once described a New York minute as 'the time it takes for the light in front of you to turn green

and the guy behind you to honk his horn'.

NO TIME TO DIE

There are hundreds of theories of how the world might end. The area of theology that looks at the final events of history even has its own name – eschatology. People have famously predicted the exact date of the apocalypse, sometimes called 'the rapture', since the first millennium with some supposed doomsday happenings proving even more unlikely than the next. Nostradamus (1503–1566), who could supposedly see future events, wrote that the 'King of Terror' would come from the sky in the seventh month of 1999. Many of Nostradamus's predictions are said to have come true, although others argue that his vague writing has meant the predictions can be moulded to fit particular events. More recently, some people believed the end of Mayan calendar on 21 December 2012 would mark the end of the world through a series of cataclysmic events. The alarming idea wasn't helped by the release of the disaster film, *2012*, showing what might happen if the prediction came true. So far, of course, no predictions about the end of the world have proven to be accurate.

NOON AND MIDNIGHT

There is a great deal of confusion as to whether noon and midnight are a.m. or p.m. In fact, they are neither; a.m. stands for the Latin words *ante meridiem* ('before the meridian') and p.m. stands for *post meridiem* ('after the meridian'). At midday, the Sun is exactly above the meridian, so it is neither *ante* nor *post*. To avoid confusion, midday or midnight can be described by adding the number twelve, to say '12 noon' or '12 midnight'. Alternatively, you might use the 24-hour clock, where midday

N

is 1200 hours (called 'twelve hundred hours')
and midnight is 0000 hours (or 'zero hundred
hours').

O'CLOCK

In the past, when people wanted to know the time, they would ask 'What of the clock?', meaning 'What time is shown on the clock?' When we say it is 'six o'clock', what we really mean is that it is 'six of the clock'.

OCTOBER REVOLUTION IN NOVEMBER

When the calendar reforms of Pope Gregory XIII were announced in 1582, not all countries adopted the new-style calendar at the same time, preferring to continue use of the Julian calendar. Italy and Portugal were among the first to make the change, Hungary followed in 1587, Denmark in 1700, Japan in 1873 and Greece in 1923. Owing to this time lag, it was often the case that the date in some countries could differ from that in others by ten days or more. Russia did not adopt the new style calendar until 1918. It is a curious fact that the famous 'October Revolution', which took place in Russia in October 1917 based on the Julian calendar in use there at the time, actually took place in November 1917 according to the Gregorian calendar, then kept in most of the rest of the world.

ONE MISSISSIPPI

Counting seconds without a clock can be a tricky business. One supposedly helpful technique is to say the word 'Mississippi' after each number – 'one Mississippi, two Mississippi, three Mississippi', etc. The time taken to say the word should equal the length of one second, although as proven in the *Friends* episode *The One with Ross's Tan*, this isn't too accurate. It is a rough indicator at best but if you say the word too quickly or too slowly, you can find yourself way off the mark, or with a ridiculously dark fake tan!

ONE SMALL STEP

How do you tell the time on the Moon? You need a pretty impressive watch! The Omega Speedmaster is designed to be shock-proof, water-resistant and able to withstand 12 g of acceleration. No wonder then that it was chosen by NASA for Apollo astronauts to wear on missions, including the famous Apollo 11 moon-landing mission in 1969. Although Neil

Armstrong was the first man on the Moon, he left his watch inside the lunar module, so it was Buzz Aldrin's that became the first to be worn on the Moon. Buzz commented on the apparent redundancy of the 'moon watch' in his book *Return to Earth*. He wrote that 'few things are less necessary when walking around on the Moon than knowing what time it is in Houston, Texas'. But the Apollo 13 astronauts may disagree after the watch helped save their lives. It was used to accurately time manual burns (a type of manoeuvre) that ensured their safe return to Earth following a system failure.

OUT OF TIME

Over a period of time, plants and animals may become extinct and therefore no longer exist: their time has literally, come to an end.

The last dodo, for example, died in the late 1600s, probably killed for food by sailors. The dodo was a bird which could not fly. It lived on the island of Mauritius in the Indian Ocean. The *Calvaria Major* tree is also found on this island. It has seeds so hard that they cannot germinate by themselves, but need to be cracked somehow before they can grow. When the dodo lived on the island, it ate these seeds and digested the outer layer, which prepared the seeds for sprouting as they left the bird's body. When the last dodo died, there were no animals left to perform this function. Today, the tree also faces extinction as there are only a dozen left on the island, although this theory is still widely debated.

THROUGH THE AGES

Time periods can be divided broadly into *prehistorical* (before history began to be recorded) and *historical* eras (when written records began to be kept). Here is a summary of the key time periods throughout history. ☞

STONE AGE
(3 MILLION YEARS AGO–3000 BCE)
Early humans lived in small nomadic groups and used tools made from stone. Over time they started to domesticate animals and create cave art.

BRONZE AGE
(3000–1300 BCE)
Cultures in Europe and Asia created bronze, an alloy of copper and tin. It became a vital material for making tools and weapons.

IRON AGE
(1200-230 BCE)
A period of economic development as people traded objects made from iron. The invention of steel provided them with stronger tools and weapons.

ANCIENT EGYPT
(3000-300 BCE)
A civilisation that inhabited the banks of the Nile. By 30 BCE, Egypt had fallen under the rule of the Roman Empire.

ANCIENT GREECE

(800 BCE–0 CE)

Ancient Greece is considered the birthplace of modern democracy and representative government. Other key developments include the Olympics, which was the cultural highlight of the Greek calendar for over 12 centuries, a tradition that continues to this day, of course!

ANCIENT ROME

(800 BCE–476 CE)

The Roman Empire extended throughout the majority of Europe and laid the foundations of Western civilisations. Towards the end of the Roman Empire, Christianity was adopted as its official religion, which helped it to spread across Europe.

MIDDLE AGES
(476 CE-1500 CE)

A challenging time in Europe with severe wars, plagues, religious persecution and political reform. Scholarship was limited to the Church and newly emerging universities.

THE RENAISSANCE
(1350s-1650s)

This period saw a rebirth of culture and knowledge as texts from Ancient Greece were translated into Arabic and Latin. The arts and sciences flourished as scholars and artists explored and challenged these ideas.

THE ENLIGHT-ENMENT

(1650s–1780s)

The Enlightenment is a period which saw the growth of intellectual reason, individualism and a challenge to existing religious and political structures.

THE ROMANTIC ERA

(1790s–1850s)

The Romantic Era was an artistic, literary and intellectual movement, defined by Romantic poets such as Blake, Keats, Coleridge, Wordsworth and Shelley, and other artists, composers and writers. The Romantic era was characterised by a focus on emotion and individualism, partly as a reaction to the Industrial Revolution.

INDUSTRIAL REVOLUTION

(1750s–1900)

The Industrial Revolution saw the transition to new manufacturing processes, with the shift from a largely agrarian economy to an industrial economy based on coal, steel, railways and specialisation of the workforce.

AGE OF IMPERIALISM

(C.1700–1950s)

The Age of Imperialism refers to the period of time when European powers began to conquer other countries and parts of the world. The British Empire, at its peak, covered 25% of the globe, in countries such as India, the West Indies and parts of Australasia.

INFORMATION AGE

(1971–PRESENT)

The Information Age refers to the new modern technologies which have shaped the modern world, including computers, the internet, mobile phones and social media.

THE PENDULUM

Galileo Galilei (1564–1642), the famous Italian astronomer and mathematician, was the first to suggest that a pendulum could be used to keep accurate time. Sitting in the draughty cathedral in Pisa in the late 16th century, he is supposed to have watched how a swinging lamp suspended from the ceiling kept a regular rhythm when he measured it against the

pulse in his wrist. The Dutch mathematician, Christiaan Huygens (1629–95), was the first person to use the pendulum in a mechanical clock in 1656.

PERFECT DAYS

The Bahá'i faith, which originated in the Middle East, named its days and months after qualities of their god. Weekdays, for instance, are called 'Glory', 'Beauty', 'Perfection', 'Grace', 'Justice', 'Majesty' and 'Independence'. Their calendar has 19 months, each of 19 days, with an extra 'month' of four or five days added to keep the length of the year in step with the seasons. The beginning of the day is measured from sunset and New Year's Day is on 21 March.

PHOTO FINISH

The human eye is often not fast enough to distinguish between two closely placed sporting competitors crossing a finishing line. In order to allow the judges to decide the winner, photography and film is used to capture split-second action. In the 2008 Summer Olympics, cameras which could take 3,000 images a second were introduced to improve accuracy, although some athletes can still register identical times. At the 2016 Rio Olympics, USA's Simone Manuel and Canada's Penny Oleksiak

both won the women's 100-m freestyle swimming final, finishing the race in exactly 52.70 seconds. Not only did they both receive gold medals but they also both achieved an Olympic record! Both flags were raised and both anthems were played at the medal ceremony with the silver podium being left empty. Some sporting events record finishing times to hundredths or even thousandths of seconds to get the most accurate results.

PIPS OF TIME

For over 75 years on BBC radio, the major news headlines of the day have been preceded by the 'six pips' of the Greenwich Time Signal. The signal was first broadcast in February 1924. The concept of the 'pips' was put forward by the Astronomer Royal of the day, Sir Frank Dyson (1868–1939). Soon afterwards, at an important dinner, he was offered six orange pips on a plate in celebration of the idea. Occasionally, there is a seventh pip to mark the addition of a leap second. The pips are timed from an atomic clock in the basement of Broadcasting House in London, which is in turn synchronised with the highly accurate time signal from the National Physical Laboratory.

POWER PLANTS

The fastest-growing plant is a particular species of bamboo, which has been found to grow up to 91 cm (35 in) per day, a speed of 0.00002 mph. The world's fastest-growing tree is the empress tree, or foxglove tree, which can grow 6 metres in its first year, and as much as 30 cm in 3 weeks.

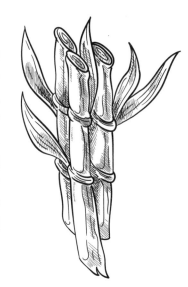

THE QUEEN IS NEVER LATE

There is a legend in the UK that it is impossible for Her Majesty the Queen to be late for the ceremony of Trooping the Colour. If she ever were to be late, it is rumoured that there is an officer with the special task of putting back the clock so that she, in fact, arrives 'on time'. Palace officials deny this. The official line is that royal engagements are so precisely timed

and everything so well-rehearsed that the Queen can never be late.

QUICK OFF THE MARK

How fast can you run? The fastest humans can finish the 100-metre sprint in under 10 seconds. Usain Bolt broke the world record in 2009 after running 100 metres in just 9.58 seconds. Some of the fastest fish are young

surgeonfish and
soldierfish that
swim in the wa-
ters of the Great
Barrier Reef and off

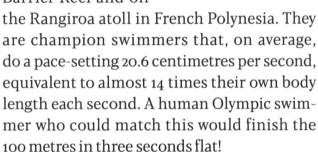

the Rangiroa atoll in French Polynesia. They
are champion swimmers that, on average,
do a pace-setting 20.6 centimetres per second,
equivalent to almost 14 times their own body
length each second. A human Olympic swim-
mer who could match this would finish the
100 metres in three seconds flat!

RECORD TIME

In athletics, changing technology has improved the precision with which sporting
events are timed. In the 1870s, winning times
were accurate to within a half-second. By
the late 1880s, events were timed to within
a quarter second, by the early 20th century,
to one-fifth of a second, and, in the 1920s, it
was possible to calculate up to one tenth of

a second. At the 1956 Olympic Games, events were timed at an astonishing one-hundredth of second.

REVOLUTION IN TIME

As a result of the French Revolution (1789–95), a new calendar was adopted in 1793. The beginning of the year coincided with the autumnal equinox on 22 September, the day when the hours of day and night are equal. This date was chosen as a symbol of the equality of all men. Each of the months was renamed from September onwards as: Vendémiaire (grape harvest), Brumaire (fog), Frimaire (frost), Nivôse (snow), Pluviôse (rain), Ventôse (wind), Germinal (budding plants), Floréal (blossom), Prairial (meadows), Messidor (harvest), Thermidor (heat) and Fructidor (fruit harvest). The English poked fun at the new French calendar, translating the months as: Wheezy, Sneezy, Freezy, Slippy, Drippy, Nippy, Showery, Flowery, Bowery, Wheaty, Heaty and Sweety.

The calendar lasted only until the end of 1805, when Napoleon quietly reintroduced the Gregorian calendar.

RHYTHM OF LIFE
All animals have a series of natural 'body clocks', which are regulated by their hormones and glands. These clocks regulate the heartbeat, waking and sleeping patterns and fluctuations in body temperature. These natural clocks are also known as the circadian

rhythms, from the Latin words 'circa' (about) and 'diem' (a day). When human beings sleep, a special growth hormone is released. So, when parents tell their children to stop watching television and go to bed, it is for a good reason. Sleep enables us to grow big and strong!

RIGHT ON TIME

Throughout the ages, people have tried to measure time more and more precisely. The earliest mechanical clocks of the 14th century were accurate to about 20 minutes per day. With the invention of the pendulum clock in the middle of the 17th century, accuracy improved incredibly to plus or minus ten seconds per day. Until the 19th century, people who owned watches had to check the accurate time by comparing the watch with a sundial. With new technology in the 1930s, quartz crystal clocks were able to keep time to about two milliseconds per day. Today, we rely on caesium fountain atomic clocks that only lose or gain one second every 158 million years. Scientists are already working on the timekeepers of the future. Called 'ion traps', these experimental timekeepers only gain or lose a second once every 33 billion years – that's over twice the age of the Universe!

SAME TIME TOMORROW?

Time loops (or temporal loops) are great fictional plot devices. Characters can appear to be stuck as they experience the same period of time over and over again. The character may do things differently or try to escape the loop, but the same events happen regardless. In the 1993 film *Groundhog Day*, TV weatherman Phil

Connors relives the annual Groundhog Day event potentially hundreds or thousands of times before finally waking up the next day, 3rd February. It isn't clear exactly how many times Phil 'loops' but it is clear how popular the film was, with the term 'groundhog day' becoming a commonly used phrase to refer to recurring events.

SEASONS

The changing seasons are caused by the tilt of the Earth on its axis as it orbits the Sun. The North Pole always points in the direction of the Pole Star. In June, the northern hemisphere of the Earth is tilted towards the Sun causing the longer, warmer days of summer. At the same time, it is winter in the southern hemisphere which is tilted away from the Sun. When the northern hemisphere is tilted away from the Sun in December and January, shorter, cooler days are the norm and it is summer in the southern hemisphere. The temperate latitudes have four seasons: spring, summer, autumn (or fall) and winter. Close to the equator, there tend to be just two seasons: wet and dry.

SHAPE OF TIME

Our senses tell us that time moves in a single direction, from the past, through the present and on to the future. For example, it seems impossible to move backwards in time, just as you cannot un-bake a cake or un-blow-out a candle. But there is also another aspect of time that we can perceive. This is the cyclical, repetitive nature of time, such as the regular pattern of the seasons and the daily rhythms of life. This regular cycle of time is echoed by Buddhist and Hindu beliefs in reincarnation, where each death signals a new life and a new beginning. In complete contrast, the Spanish surrealist artist, Salvador Dali (1904–89), created a painting called *The Triangular Hour*. What shape do you imagine when you think of time?

SLEEPING BEAUTIES

Wednesday, 2 September 1752, was a great day in the history of sleep. That evening, millions of people in Great Britain and America went to bed, not to wake up until 14 September. This amazing feat was not caused by a magical spell or a powerful sleeping drug. Instead, it was caused by the fact that the calendar had just been changed. Pope Gregory had developed a new calendar in 1582, but it was not introduced into Great Britain and North America until 1752, when it was agreed that Wednesday 2 September should be followed by Thursday 14 September.

The calendar change provoked a public outcry by those who felt confused and cheated. Some unscrupulous landlords insisted on having those 11 days' worth of rent paid, even though the actual calendar dates did not exist. In the City of London, bankers protested about the change by refusing to pay taxes on the usual due date of 25 March. They insisted on paying 11 days later, on 5 April, which remains the end of the tax year in Britain today.

SNAIL'S PACE

The common garden snail travels at a grand speed of 0.03 mph. Compare this with the one of the fastest land animals, the cheetah, who can reach speeds of around 60 mph, but there's an even faster animal still. Several birds, and a type of bat, can travel faster than the cheetah, with the fastest bird being the peregrine falcon. It can reach over 200 mph when diving!

SOUND THE ALARM

We all love a snooze, or do we? According to a YouGov survey, 41% of British people don't hit the snooze button and 21% don't even set an alarm! If you do need some help waking up, it might be better to invest in a physical alarm clock. Studies show you're more likely to oversleep if you use your smartphone as your morning wake up call.

THE SPEAKING CLOCK

Telling the time by telephone was first established at the Paris Observatory in 1905. The service proved very popular, but it was too time-consuming for observatory staff who had to answer the telephone and count the minutes and seconds out loud along with the

ticking of the clock. In February 1933, a new service was offered – *l'horloge parlante*, or 'the speaking clock'. This was a fully automated service based on similar systems available in other countries. In Britain, the speaking clock was brought into use on 24 July 1936. The current voice of the speaking clock is Alan Steadman, a local radio broadcaster from Dundee, Scotland. BT's speaking clock is one of very few left in the world today. AT&T discontinued their service in the United States in 2007.

SPLITTING THE SECOND

The femtosecond is a minuscule one-mil-lionth of a billionth of a second (10^{-15} seconds). It is the time taken for an atomic bond to break during a reaction. If time were slowed down so much that such events lasted a full second, the world record for the 100-metre sprint (just over 9.5 seconds) would be equivalent to 320 million years. Until recently, scientists said it was impossible to time such small events, but now they can measure them with pulses of laser light.

A STAR WHICH TELLS THE TIME

Pulsars are the remnants of exploded stars that condense into rapidly rotating cores and emit radio waves. It is similar to the rotating beam on a lighthouse, appearing to flicker on and off with amazing precision. The first pulsar was discovered in 1967 by Dame Susan Jocelyn Bell Burnell when she was a postgraduate student. She and her advisor, Antony Hewish, had no idea what these signals were, so they called them LGM, or 'Little Green Men'. More than 2,000 have been discovered since. The fastest known pulsar rotates over 700 times per second. Astronomers continue to use these celestial timekeepers to detect gravitational waves, track distant space probes and monitor the accuracy of atomic clocks on Earth.

STAYING IN SYNC

For astronauts on the International Space Station, sunrise and sunset occur every 45 minutes as they orbit the Earth. Our bodies rely on sunlight to keep in sync with our natural circadian rhythm of 24 hours so many astronauts quickly feel jetlagged. Scientists are investigating how this affects astronauts' mental and physical health on long-duration space missions.

STOATS IN TECHNICOLOUR COATS

There are a number of animals whose coats change colour with the changing seasons. In most cases, the new coat provides better camouflage against the changing scenery. In summer, the stoat has a brown coat. As winter approaches, the coat changes to the white fur known as 'ermine'. The whiteness of the coat depends upon the coolness of the winter temperature, so a pure white coat is produced in the snowy north, but a mottled coat in the warmer south. This change makes it easier for the stoat and other animals to escape predators.

STOLEN TIME

A klepsydra is a clock which measures intervals of time against the flow of water. The name comes from the Greek words meaning 'water thief'. Klepsydra were used by the ancient Babylonians, Egyptians and Chinese. The earliest water clocks simply consisted of a bowl with a hole in the bottom which would be placed in deep water. As water gushed through the hole, the bowl would eventually sink and this could be used to time a short event such as a cock fight.

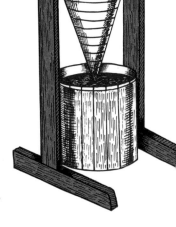

STOP ALL THE CLOCKS

Clocks stopping has often been associated with the image of death. When clocks and watches needed to be wound by their owners, they would indeed stop (eventually) after

they died as they no lon-
ger had a source of power.
Made famous by the film
*Four Weddings and a Fu-
neral*, W.H. Auden's poem
'Funeral Blues' starts with
the well-known words
'stop all the clocks' as
it describes the feeling
of losing a loved one.

Charles Dickens used the stopping of clocks
to signify the death of love in his novel *Great
Expectations.* All the clocks and watches in
the bitter Miss Havisham's house are stopped
at 8.40, the exact time that she was jilted by her
husband-to-be.

STRAWBERRY MOONS

Because of its distinctive phases the Moon
has been used as a way of measuring time by
many different cultures. The indigenous tribes
of North America name the months after

seasonal events, each 'Moon' signifying the activities associated with a particular time of year. The Natchez tribe's year, for example, starts in March with the months called Deer Moon, Strawberry Moon, Little Corn Moon, Watermelon Moon, Peaches Moon, Mulberries Moon, Maize Moon, Turkey Moon, Bison Moon, Bear Moon, Cold Meal Moon, Chestnuts Moon and Nuts Moon.

SUMMER NIGHTS

Olivia Newton John and John Travolta sung about the joys of summer nights but there's an important reason why they are so short. The summer solstice is often referred to as the longest day of the year – the number of hours of daylight are at their maximum, while the number of hours of night are at their minimum. It is also when the Sun reaches its highest position in the sky. In the northern hemisphere this solstice is in June and in the southern hemisphere it is in December. In the Arctic and Antarctic Circles there is continuous day-light around the summer solstice.

SUNDIALS AND SHADOWS

One of the simplest ways to measure time is with a sundial. The dial makes use of light and shadow. The earliest sundials were simply sticks set in the ground. As the Sun appears to move across the sky from east to west, the shadow cast by the stick moves across the ground as well. The length of the shadow and its direction can provide information about the time of day. Time told by a sundial is known as 'solar time' or 'local time'.

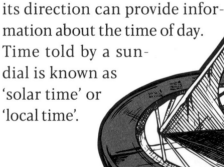

THICK AS A PLANCK

The smallest division of time is called 'Planck time'. It is named after the physicist Max Planck (1858–1947). One minute contains 600,000,000,000,000,000,000,000,000,000,000,000,000,000,000,000 Planck times. At the other end of the spectrum, the word for the longest measurement of time is *kalpa*, which is Hindi for 4.32 million years. This is the period of time that Hindus believe each reincarnation of the world lasts.

THINK FAST

How fast can you think? According to scientists, 220 miles per hour is the top speed at which signals travel through the brain. These signals direct all our thoughts and actions.

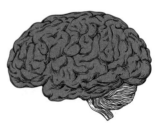

TICKING ALONG

Timekeeping devices have certainly evolved from their early days as sundials, candles or water clocks. The first mechanical clocks were invented around the beginning of the 14th century and some still exist today, although there is some debate over which one is the oldest surviving clock. It has been said that the iron-framed clock in Salisbury Cathedral is the oldest working clock in the world, supposedly dating back to 1386, although this is unconfirmed. The Backhaus clock in Forchtenberg in Germany is the oldest clock with a proofed engraved date, 1463.

TIME BALLS

When the time ball was erected at the Royal Observatory, Greenwich, in 1833, the whole area below the Observatory around the River Thames consisted of working docks and wharves. Navigators on the river looked to the time ball for a daily time signal to set the chronometers on board ship. During the 1820s, there were many different types of time signal made from the shore, including firing guns or dipping flags. The first time ball in England was probably set up in Portsmouth in 1829. At Greenwich, the time ball has always been dropped at 1 p.m. rather than at 12 noon. This is because at noon the astronomers were occupied taking observations of the Sun crossing the meridian.

TIME CAPSULES

A time capsule is a vessel designed to hold documents, objects and messages for future generations to discover. If you decide to create a time capsule, remember to think carefully about what should be included and for how many years the contents need to survive. Newspaper quickly yellows and crumbles, and although a recording on a memory stick might be an obvious choice as a record of everyday life today, the technology used to play it may have disappeared by the time the capsule is opened. Most importantly, remember to leave precise instructions on where to find your time capsule if it is ever to be rediscovered!

TIME FLIES

When you hear an aeroplane and look up in the direction of the sound, you may wonder why you often cannot spot it straight away.

The aeroplane usually seems to be flying a good distance ahead of the roar of its engines. This is because sound travels about a million times more slowly than light. Sound travels at about 750 miles per hour while light travels at around 186,000 miles per second. For an aeroplane flying at 30,000 feet, it takes about 30 seconds for the sound to reach the Earth whereas light gets here almost immediately. So a plane travelling at 600 miles per hour has already flown another 5 miles by the time you look up to see where the noise came from.

TIME FOR THE WORLD

At the International Meridian Conference held in Washington DC in 1884, 25 countries met to decide which of the many national meridians would be recognised as the Prime Meridian of the world. When it came to the vote, 22 countries voted for the meridian passing through the Royal Observatory in Greenwich. The conference delegates approved the creation of a 'Universal Day' that began at midnight in Greenwich. People started to describe their local time zone as the number of hours before or after Greenwich Mean Time (GMT). One of the countries which abstained from voting for the Greenwich Meridian was France. Paris had an important observatory of its own and for many years France insisted on referring to Greenwich Mean Time (GMT) as 'Paris Mean Time diminished by 9 minutes 21 seconds'. In fact, this is equal to GMT but avoids use of the word 'Greenwich'.

TIME OF THEIR LIVES

The average length of an animal's life depends on many things such as the size of the animal and metabolic rate. Bigger animals tend to live for longer. This is certainly true if you compare the common mayfly that lives for just 24 hours to the bowhead whale at a staggering 200 years, but it isn't always the case. A queen ant can live for decades, whereas a much bigger mouse only lives for around 4 years. Biologists say that the lifespan of a species is about one billion heartbeats.

TIME WITH CATS

It is claimed that the Austrian neurologist Sigmund Freud once said 'time spent with cats is never wasted', but it seems cats are the ones who really know how to waste time. On average cats sleep for around 15 hours every day. That's more than 60% of their lives dedicated to sleep! The main reason for this is energy conservation, but one person who might not

care is Freud himself. There's no evidence that he ever came up with the feline-loving quote, especially as he did once write to a friend 'I, as is well known, do not like cats'.

TIME ZONES

The time zone system we use today was devised by an American professor named Charles Ferdinand Dowd (1825–1904). The United States is so wide that there is a difference of several hours of local time between the east and west coasts. In 1870, Dowd suggested dividing this distance into an equal pattern of zones that were each 15° of longitude wide. The same time is kept by all towns located within a single zone. The zone to the east is exactly one hour ahead and the one to the west is exactly one hour behind. In theory, we could divide the Earth's daily turn on its axis into 24 segments with each segment covering 1 hour of rotation (15° longitude). Everyone situated within the same segment would use the same number of hours before or after GMT as their local standard time. In reality, people have created their own geographical regions (time zones) where everyone follows the same time. Over the past century, governments have chosen (and sometimes changed their minds!) about the best time zone to suit their political, social and economic needs. This means we now have over 35 time zones worldwide.

TIME-LAPSE PHOTOGRAPHY

Time-lapse photographs can give the impression of being able to slow down events to allow us to see the individual beats of a hummingbird's wings, or the splash created by a single drop of water. Events can also appear to be speeded up to allow us to see

an image of a plant growing over several days or the path of the stars wheeling across the sky throughout the night, in a matter of seconds. We probably see time-lapse as a relatively new invention, but it dates from the 1870s. The invention is credited to the photographer Eadweard Muybridge (1830–1904) who was asked to prove whether or not racehorses' hooves are ever simultaneously in the air when running. He set up a series of cameras which were set off by a tripwire when the horse ran past. The photos were then put together to create a series of images showing the horse running, and yes, all hooves were off the ground for an instant!

TIMETABLE NIGHTMARES

Until the middle of the 19th century, there were no universally accepted rules for keeping time. Most people tended to use a sundial to tell the time. As a sundial gives 'local' or 'solar' time, this means that it also gives different times in towns that are east or west of each other. Since the Sun appears to rise in the east and set in the west, it passes over places in the east sooner than places in the west. This means that sundials in the east show a time that can be significantly ahead of those in the west. For example, when it is 12 noon on a sundial in Greenwich, it is already 12.05 p.m. in Norwich, but still only 11.48 a.m. in Dundee. This difference in local times caused a real problem for the first railway timetables and often resulted in passengers missing their trains. In November 1840, the Great Western Railway (GWR) ordered that 'London Time' should be kept at all railway stations. Strictly speaking, 'London Time' was defined as the time measured from the meridian passing through St Paul's Cathedral, located 23.1 seconds west ('slow') of Greenwich. Yet most public clocks were set to Greenwich Mean Time (GMT) so the 'London Time' shown on a railway clock was actually GMT. Many other railway companies followed suit until, in 1880, GMT was adopted as legal time throughout Great Britain. At the same time, it was adopted as the official time for railway timetables.

TURN BACK TIME

Many countries advance their clocks by an hour in the summer months as a way of making the evening hours lighter for longer. In the United Kingdom, this is known as British Summer Time (BST) and in other countries as Daylight Saving Time (DST). The idea of advancing clocks to conserve daylight was the brainchild of William Willett (1856–1915), a London builder living in the south-east suburbs, who loved the outdoors. On an early morning horse ride, he noticed how many curtains were still drawn blocking out daylight, and came up with the idea of moving the clocks forward for summertime. BST was first introduced in 1916 during the First World War, when it was used as a way to increase the daylight hours available for working in factories. This helped save fuel because every electric light in the nation was switched off for an extra hour each evening. There is a saying which helps people to remember which way to turn the hands of the clock: 'spring forward' (end of March); 'fall back' (end of October).

TURN OF THE CENTURY

Even though a century is defined as a period of 100 years, there are two centuries that are different in length. The 1st century BC was longer by 90 days due to the changes that Julius Caesar made when he introduced his Julian calendar in 45 BC. In many Catholic countries, the 16th century was ten days shorter than normal because of the introduction of the Gregorian calendar. In Britain and America, however, it was the 17th century that was shorter by 11 days for the same reason.

UNEQUAL HOURS

We are used to having hours of an equal length, but this has not always been the case. Before the emergence of mechanical clocks in the 13th century, many people kept what were known as 'unequal' or 'seasonal' hours which changed length according to whether it was day or night or according to the season. To calculate unequal hours, the day was divided into two halves: from sunset to sunrise (the hours of darkness) and from sunrise to sunset (the hours of daylight). Each day consisted of twelve hours of darkness and twelve hours of daylight. The problem with this system is that the length of daylight and darkness changes throughout the year. In summer, the daylight

hours are longer than the night, and in winter the night hours are longer than the day. There are only two days out of the whole year when the hours of the day and the hours of the night are equal. These occur at the spring and autumn equinox, around 21 March and 22 September. One country that still keeps unequal hours today is Ethiopia, but the people there are lucky because, living close to the equator, the lengths of the nights and days are nearly equal throughout the year!

VERY FAST DAYS

A day equals the time it takes for the Earth to make one complete turn on its axis, approximately 24 hours. As the Earth revolves, the side facing the Sun is light and the other side is in darkness. Similarly, all of the other planets revolve at different speeds and their days are longer or shorter than a day on Earth. A day on Neptune lasts only 16.1 Earth hours. The shortest days in our solar system are found

on minor planets known as 'fast rotators'. They can spin in as little as 12 seconds! One such asteroid, called 1998 KY26, is just 30 metres across and spins once every 10.7 minutes. Imagine having to cram your daily routine - brushing your teeth, going to school or work, eating breakfast, lunch and dinner, sleeping then waking up again – all into ten minutes!

VERY LONG DAYS

Venus rotates on its axis so slowly that a Venusian day (from sunrise to sunrise) lasts about 243 Earth days. Venus is very close to the Sun, which it orbits very quickly. A year is the time it takes for objects to complete one orbit around the Sun, so a year on Venus lasts just 224.7 Earth days. This means that the days on Venus are actually longer than its years. This would mean that, by the time an Earthling baby reaches its first birthday, a baby born simultaneously on Venus would already be two years old.

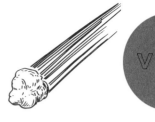

How long is a day or a year on the other planets in the Solar System?

Planet	Length of a day	Length of a year
Mercury	59 Earth days	88 Earth days
Venus	243 Earth days	225 Earth days
Mars	Just over 1 Earth day	687 Earth days
Jupiter	10 hours	Almost 12 Earth years
Saturn	Almost 11 hours	29.5 Earth years
Uranus	17 hours	84 Earth years
Neptune	16 hours	Almost 165 Earth years

WEEKS AND PLANETS

The week is a period of seven days, the time taken by the Jewish and Christian God to create the world. The number and names of the days of the week are based on what ancient people thought of as the 'seven planets' which they believed to orbit the Earth: the Sun, Moon, Mars, Mercury, Jupiter, Venus and Saturn. In those days, they counted the Sun and Moon as planets. The names for the days

Thank the Gods it's Freitag

Latin	Old English	English
Dies Lunae	Moon's Day	Monday
Dies Martis	Tiw's Day	Tuesday
Dies Mercurii	Woden's Day	Wednesday
Dies Jovis	Thor's Day	Thursday
Dies Veneris	Frigg's Day	Friday
Dies Saturni	Seterne's Day	Saturday
Dies Solis	Sun's Day	Sunday

of the week in different languages are given in the table below. Many European languages adopted the names of the day from the Latin, but the English language adopted some of the names from Scandinavian mythological gods associated with the planets. Tuesday is named after Tiw, god of war; Wednesday is named after Woden, god of the underworld; Thursday is named after Thor, god of thunder; and Friday after Frigg, goddess of love. The word

German	*French*	*Italian*	*Spanish*
Montag	Lundì	Lunedì	Lunes
Dienstag	Mardi	Martedì	Martes
Mittwoch	Mercredi	Mercoledì	Miércoles
Donnerstag	Jeudi	Giovedì	Jueves
Freitag	Vendredi	Venerdì	Viernes
Samstag	Samedi	Sabato	Sábado
Sonntag	Dimanche	Domenica	Domingo

'fortnight' is an abbreviation for 'fourteen nights'.

WHO WANTS TO LIVE FOREVER?

People have been obsessed with achieving eternal life for thousands of years. The Ancient Greeks attempted to create a philosopher's stone that would let humans live forever. Sadly the idea only lives on in fictional realms such as Harry Potter's wizarding world. But apparently immortality is closer than we think. Scientists claim that if we can make it to the year 2050 we could stand a chance of living forever thanks to constantly improving technology. Some options could include renewing body parts or becoming an android, but it might cost you an arm and a leg in more ways than one!

WINTER IS COMING

Everyone knows when winter is on its way – the days get shorter and the nights get longer. In the northern hemisphere, there comes a point where a 24-hour period has the fewest daylight hours of the year. This is known as the winter solstice. People refer to it as the shortest day of the year, or the longest night of the year. In London, the shortest day is just under 8 hours. The date of the winter solstice changes, but it is usually around 21 or 22 December

in the northern hemisphere. It's the opposite story in the southern hemisphere where the *summer* solstice occurs on the same date. The situation is reversed again around 21 June. This is because the Earth's North Pole is pointing away from the Sun (making it colder) and the South Pole is pointing towards the Sun (making it warmer).

WORKING 9 TO 5

The '9 to 5' is what most people consider to be an average work day, but not everyone gets to sing the famous Dolly Parton song! Workers in Mexico spend the longest time at work, clocking up 2,255 hour each year – that's about a 43-hour work week! In Germany, workers spend just 1,363 hours at work.

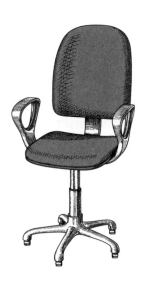

An initiative set up in 2018 is mostly to thank for this as it gave German people the right to work a maximum of 28 hours per week. The difference can also be down to vastly different economic climates and cultural outlooks across the globe.

WRIST WATCHES

Watches were traditionally worn by men as a 'fob-watch' on the end of a chain which could be tucked into a waistcoat pocket. The earliest wristwatch was simply a fob-watch placed in a leather case and strapped to the wrist. Interestingly, early wrist or bracelet watches were worn mainly by women. They were advertised as 'liberating' devices, which allowed them to know the time while cycling or playing tennis. The practical nature of wrist watches made them popular with men during the First World War.

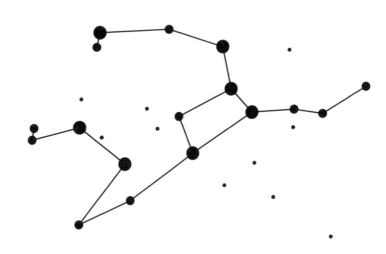

WRITTEN IN THE STARS

The stars of the night sky act as a giant clock, making one complete revolution around the Pole Star in just under 24 hours. The positions of the stars may be read like the hour hand on a clock with an old scientific instrument called a nocturnal. The name of this instrument comes from the Latin word *nox*, meaning 'night'. Once the instrument is set up with the correct date, the index arm of the nocturnal can be aligned with Ursa Major or Ursa Minor, the constellations of the Great Bear or the Little Bear, indicating the time on a numbered scale. In the dark, the hour can be discovered by feeling and counting the little teeth cut into the scale.

XIUHTECUHTLI

Xiuhtecuhtli is an ancient Mexican god of fire and time and is sometimes depicted with a furnace on his head. At the end of every 52 years, all fires were extinguished in his honour and a fresh one kindled on the chest of a human sacrifice in order to keep time moving.

XTRA TIME

Stoppage or injury time in football matches, added at the end to allow for stoppages in play, can often decide the final result of a game. The outcomes of some of the most

memorable sporting occasions have been decided in the last few minutes or, in some cases, seconds of the event. One of the most remarkable stoppage-time goals of recent years came from Ben Watson, who scored for Wigan in the 91st minute of the 2013 FA Cup final versus Manchester City. It seemed miraculous, not only because Manchester City were former FA Cup and Premier League winners, but also because Wigan were about to be relegated from the Premier League. In fact they became the first club to lift the FA Cup and get relegated in the same season. Manchester City's manager, Roberto Mancini, on the other hand, was fired two days later, proving how crucial these extra minutes can be to results and careers!

YEAR OF CONFUSION

The modern calendar is derived from one that was first developed in 46 BC, when Julius Caesar was the leader of the Roman Empire. This calendar is called the Julian calendar.

The length of the year was fixed at 365 days and every fourth year had an extra day added so that it was 366 days long. This is known as a leap year. At the beginning of 46 BC, the calendar in use in Rome was a real mess. It had slipped out of synchronisation with the stars and seasons by about three months. In order to realign the

calendar, Caesar had to add some extra days. 46 BC was an extraordinary 445 days long. Not surprisingly, the year became known as 'The Year of Confusion'.

A YEAR OF NEW YEARS

Different cultures celebrate the New Year on different dates. Many Christian countries used to start the New Year on 25 March, the feast of the Annunciation in the Christian calendar. The ancient Egyptians measured the beginning of the year by the rising of the bright star Sirius, which signalled the annual flooding of the Nile. The calendars of different countries and cultures are rarely synchronised from one year to another. For example, for the year 2000, the Chinese New Year fell on the 5th February, the Sikh New Year on the 14th March, the Hindu New Year was on the 13th April, the Ethiopian New Year was on the 11th September and the Jewish New Year was on the 30th September, according to the Gregorian calendar.

YEAR OF THE HORSE

No matter what day they are born on, all racehorses have their birthdays on the same day. In the northern hemisphere that's 1 January, while in the southern hemisphere it's 1 August. This means that a horse born on 31 December or 31 July, depending on where they live, becomes one year old the very next day. Horse breeders do everything they can to make sure their horses are born at the right time of the year – a younger horse will have less training and won't compete well against older horses that are officially considered to be the same age!

ZERO LONGITUDE

Longitude is essential to the measurement of time around the world. The lines of longitude and latitude are familiar from atlases and maps. They create an imaginary grid which can be used to pinpoint any location on the surface of the Earth. What many people do not realise is that the measurement of longitude (position east or west) is directly related to the measurement of time. The Earth takes 24 hours to complete one full revolution of 360°. This means that in one hour, the Earth revolves one twenty-fourth of a spin, or 15°, or that in 4 minutes it revolves 1°, and so on. This was an important fact for early explorers who found it impossible to navigate a ship while out of sight of land. To calculate longitude at sea, a navigator needs to know two things: what time is it on board ship, which can be measured using the Sun by day or the stars at night, and what time is it back home. The time difference converts into a longitude difference. Knowing the time back home was the biggest problem because there was no clock available which could cope with keeping accurate time on a rolling ship. Many ideas were put forward for solving the 'longitude problem' which was the greatest scientific puzzle of the eighteenth century.

ZULU TIME

Some international organisations use military time zones to communicate time. Imagine a map of the world split into every major time zone and each one is allocated a different letter of the English alphabet. It's easier and less confusing to describe each time zone by letter, rather than by saying 'X hours behind/after Greenwich'. The global sequence begins with the letter A (GMT+1 hour) and ends with Z (GMT). In the phonetic alphabet, the letter Z is referred to as 'Zulu'. This is why GMT is sometimes known as Zulu Time. The letter J, or 'Juliet', isn't used as some alphabets do not have a 'J' and it can be confused with other letters.

TIME TO TALK

Did you know that the word 'time' is the most frequently used noun in the English language?

A devil of a time

A QUESTION OF TIME

A RACE AGAINST TIME

IT'S ABOUT TIME

Quality time

AHEAD OF ONE'S TIME

A stitch in time '... saves nine'

ALL IN GOOD TIME

FOR OLD TIMES' SAKE

All the time in the world

Bide

one's

time

ALL
TIME
LOW

AS TIME GOES BY

FALL ON Crunch time

HARD TIMES FACE

Down time TIME

BEHIND

THE TIMES

FOR THE TIME BEING

Have time on one's side

GIVE SOMEONE

A HARD TIME

JUST IN TIME

From time to time

Pass the time of day

Good times

In next to no time

HIGH TIME

Have a time of it

HAVE A WHALE OF A TIME

HIT THE BIG TIME

In good time

IN LESS THAN NO TIME

In *In the right place*
at the right time

your

own

time *IN THE NICK OF TIME*

IN RECORD TIME

In the fullness of time

IT'S
HIGH
TIME

MOVE
WITH
THE
TIMES

LOSE TRACK
OF TIME

Once in a lifetime

LIVING ON TIME

ON One more

BORROWED time

TIME

KEEP

TIME

Once
upon
a
time

Long time no see

ON **MAKE GOOD**
BORROWED **TIME**
TIME *Time heals*
No   *all wounds*
time   **MAKE UP**
like   **FOR LOST**
the   **TIME**
present
OUT OF One
        step
TIME   at a
        time

SCREEN *TIME*

THE SANDS OF TIME

Time of

THERE'S A TIME AND A PLACE

one's

life Old

Take one day at a time

100 before your time

Time after time

SIGN OF THE TIMES

Time is money

TIME AND TIME AGAIN

TIME IS OF THE ESSENCE

Time flies when you're having fun

Play for time

Time off

TIME OUT

Time bomb

TIME AND TIDE WAITS FOR NO MAN

Only a matter of time

TURN BACK THE HANDS OF TIME

To buy time

TO KILL TIME

ONLY TIME WILL TELL

To call time on

Pressed for time

WITHSTAND THE TEST OF TIME

The time has come

RACE AGAINST

To do time TIME

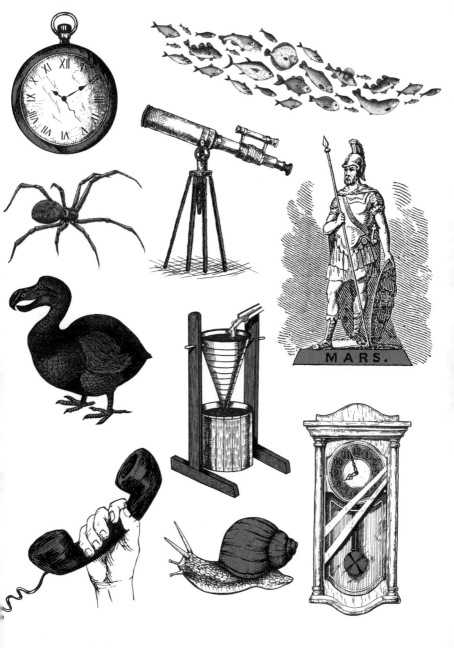

ABOUT THE ROYAL OBSERVATORY GREENWICH

The historic Royal Observatory has stood atop Greenwich Hill since 1675, and documents centuries of astronomical observation and timekeeping. It is truly the home of space and time, with the world-famous Greenwich Meridian Line, awe-inspiring astronomy and the Peter Harrison Planetarium. The Royal Observatory is the perfect place to explore the Universe with the help of our very own team of astronomers. Find out more about the site, book a planetarium show, or join one of our workshops or courses online at rmg.co.uk.